CHAPTER 1

INTRODUCTION

1.1 GENERAL

In pneumatic system compressed air enters into the tube at one end of the piston and, hence impact force on the piston. Consequently, the piston becomes displaced (moved) by the compressed air expanding in an attempt to reach atmospheric pressure.

1.2 PROBLEM STATEMENT

Generally, coconuts are de-husked manually using either a machete or a spike. These methods require skilled labor and are tiring to use. Attempts made so far in the development of De-husking tools have been only partially successful and not effective in replacing manual methods. The reasons quoted for the failure of these tools include unsatisfactory and incomplete de-husking, breakage of the coconut shell while de-husking, spoilage of useful coir, greater effort needed than manual methods, etc.

The present work involved the design, development and testing of a coconut de-husker which overcomes the drawbacks of the previously reported implements. The design and developmental stages called for a closer look at the magnitude and direction of the de-husking forces and their generation mechanisms. Details of a simple, sturdy and efficient manual de-husker unit, financially beneficial to labor and producers are given here. Comparative assessment of this unit in relation to those reported in the literature is provided. Test results and assessment of the present unit in both laboratory and field conditions are also reported. Safety aspects are incorporated. The unit can de-husk about 70 coconuts per hour compared with about 40

1

nuts per hour from a skilled worker using the spike method. It can be operated by unskilled laborers. Cost benefit analysis indicates that it should be commercially viable.

1.3 SOLUTION

The above defined problems of coconut De-husking can be reduced by using this pneumatic De-husking system. Such type of De-husking machine is low economical and give the best performance.

1.4 PNEUMATIC CYLINDER

Pneumatic cylinder is a mechanical device that produce the forces and power air or compressed gas. They carry out a lot of function in automotive, electronics, packaging industries, various establishment, and even in households. They are used to control pressure in air brakes, help car engines push the piston enabling the tire to spin, and allow certain types of door to close, among other functions. The pneumatic cylinder may have a variety of uses, but their basic function does not change. Their basic function does not change. They convert air pressure in to linear motion by attachment to a metal piston, pushed back and forth by columns of air.

Fig. 1.1Pneumatic cylinder

1.5 FAIL SAFE MECHANISMS

Pneumatic systems are often found in setting where even rare and brief system failure is unacceptable. In such situations the de husking can sometimes serve as asafety mechanism in case of loss of air supply(or its pressure falling) and, thus remedy or abate any damage arising in such a situation. Due to the leakage of air from input or output reduces the pressure and so the desired output.

1.6 WORKING OF PNEUMATIC CYLINDER

Pneumatic cylinders come in many shapes and sizes, which typically range from 2.5mm to 400mm diameter air cylinder. Some pneumatic cylinders are 1000mm diameter, instead for use in place of hydraulics. The most common pneumatic cylinder is tube shaped and contains a shaft a rod, and a plunger. The rod fit in to the shaft of the cylinder, which can be move in and out. The plunger is connected to the end of the rod, and its purpose is to take the impact of the air pressure that flows with in the pneumatic cylinder.

In order to perform, pneumatic cylinders deliver a force by transforming the potential energy of compressed gas in to kinetic energy, which is a scalar quantity, and therefore has magnitude but no direction. That direction is attained by manual or electrical value that controls the direction of movement. These direction control valves are maneuverbyelectric solenoid or hand lever to keep adjustable travel rates. When the compressed gas expands due to the force created by greater pressure than the atmospheric pressure, the air expansion drives a piston to move in a preferred direction. Some cylinder has cushions on one or both port ends. These cushions constrain airflow to reduce the motion of travel to a lower speed. It is as shown in the Fig 1.2

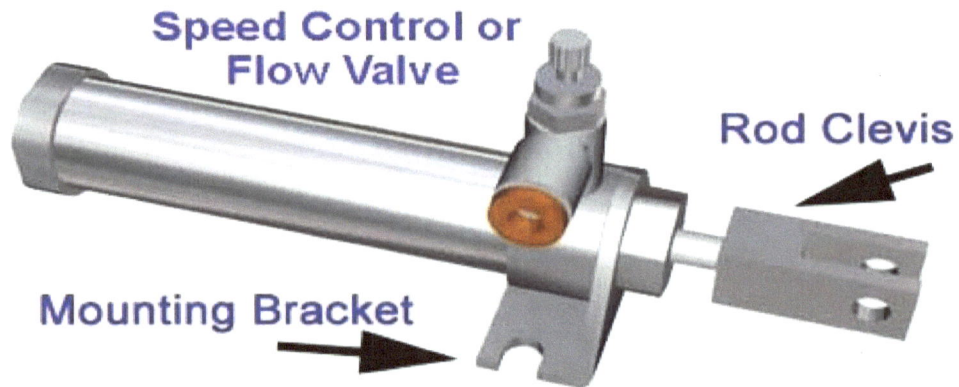

Fig.1.2 pneumatic cylinder

CHAPTER 2

LITERATURE SURVEY

2.1 SIMILAR PRODUCTS AVAILABLE IN MARKET

TYPE 1

This procedure required skilled labor these are successfully, practically, and more effective method that are available in market.

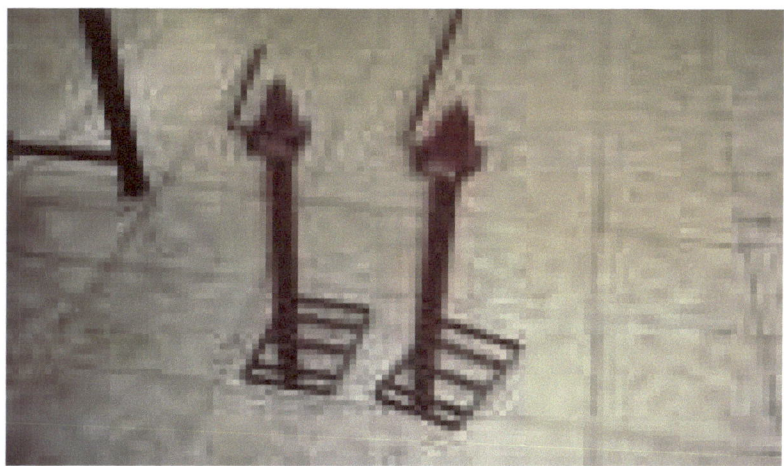

Fig. 2.1 type 1 de-husking machine

Model	Manual De- husking tool
Tool construction	Iron
Capacity	1 coconut per 1min
Tool weight	3kg
Tool size	1.5m

Table 2.1 specification of type 1de-husking machine

TYPE 2

It is an oldest method that we used in our house. The time taken to de-husk a coconut of this Machine is high

FIG 2.2 TYPE 2 DEHUSKING TOOL SPECIFICATIONS

Fig. 2.2 type2 de-husking machine

Model	De husking using iron rod
Tool construction	Iron rod
Capacity	1 coconut per 2 min
Tool weight	5kg
Tool size	2m

Table 2.2 specification of type 2 de-husking machine

TYPE 3

A better grip on the coconut is provided by the iron plate, which acts as the stopper that prevents the nut to slip away vertically. But the problem in this machine is that the hands may get damaged because the worker has to hold the coconut in his hand during de-husking.The process need electrician during the de-husking.The process need electrical.

Fig. 2.3 type 3 de-husking machine

Model	Two blade de husking machine
Machine construction	Iron rod & motor
Capacity	2coconut per 1min
Machine weight	30kg
Machine size	3.5m *

Table 2.3 specification of type 3de-husking machine

2.2 SOME OF THE PATENT PRODUCTS

Cited Patent	Filing date	Issue date	Original Assignee	Title
US801593	28 Jul 1904	10 Oct 1905		CASEIN-MANGLING MACHINE
US807551	26 jul1904	19 dec 1905		POD-OPENING MACHINE
US1554516	17 jul1920	22 sep 1925		METHOD OF EXTRACTING COCONUT MEAT
US1808745	24 may1928	9 jan 1931		PINON NUT SHELLER
US3605834	5 jan 1969	20 sep 1971		COCONUT BREAKING MACHINE
US4347260	27 jan 1980	31 aug 1982	Alf Hannafor& Co. Pty. Limited	METHODS OF SEPARATING ALMOND KERNELS FROM ALMONDS WITH SHELLS

Table 2.4 Patent Products

CHAPTER 3

PRINCIPLE OF OPERATION

3.1 BLOCK DIAGRAM

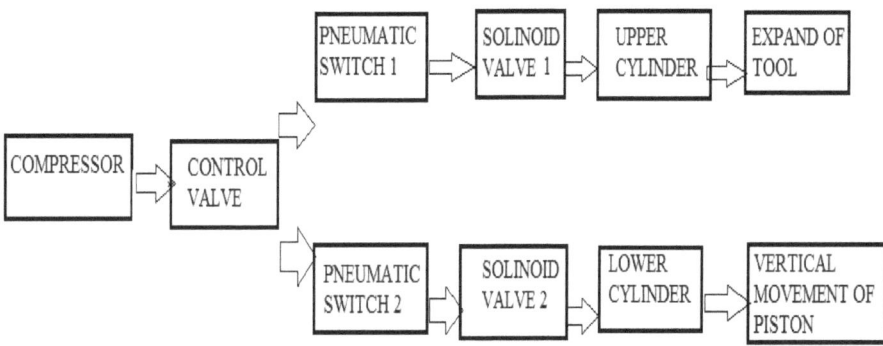

Fig.3.1 block diagram

3.2 DESCRIPTION

The basic principle is convert the pneumatic energy into mechanical energy. Here the high pressure air from the compressor is controlled by control valve then, pneumatic switch give the electrical control to the system. Solenoid valve give required control as per the electrical signal from the pneumatic switch. The cylinder can expand and retract as per the signal from the solenoid valve.

CHAPTER 4

COMPONENTS DESCRIPTION

4.1 PNEUMATIC CYLINDER

Linear actuators those are available in thousands of different configurations. Cylinders refer to devices with pistons of various diameters and strokes of various lengths. They are most commonly specified as single acting (powered in one direction) or double acting (powered in both directions).Single acting spring return cylinders are more economical with respect to air consumption. The pneumatic cylinders supplied in the GEARS-IDS Invention and Design System is single acting, spring return valves.

4.2 TYPES

Although pneumatic cylinders will vary in appearance, size and function, they generally fall into one of the specific categories shown below. However there are also numerous other types of pneumatic cylinder available, many of which are designed to fulfill specific and specialized functions.

1. Single acting

2. Double acting

4.2.1 SINGLE ACTING CYLINDERS

Single acting cylinder (SAC) use the pressure imparted by compressed air to create a driving force in one direction (usually out), and a spring to return to the "home" position. The diagrammatic representation for the single acting cylinder as shown in the Fig 4.1

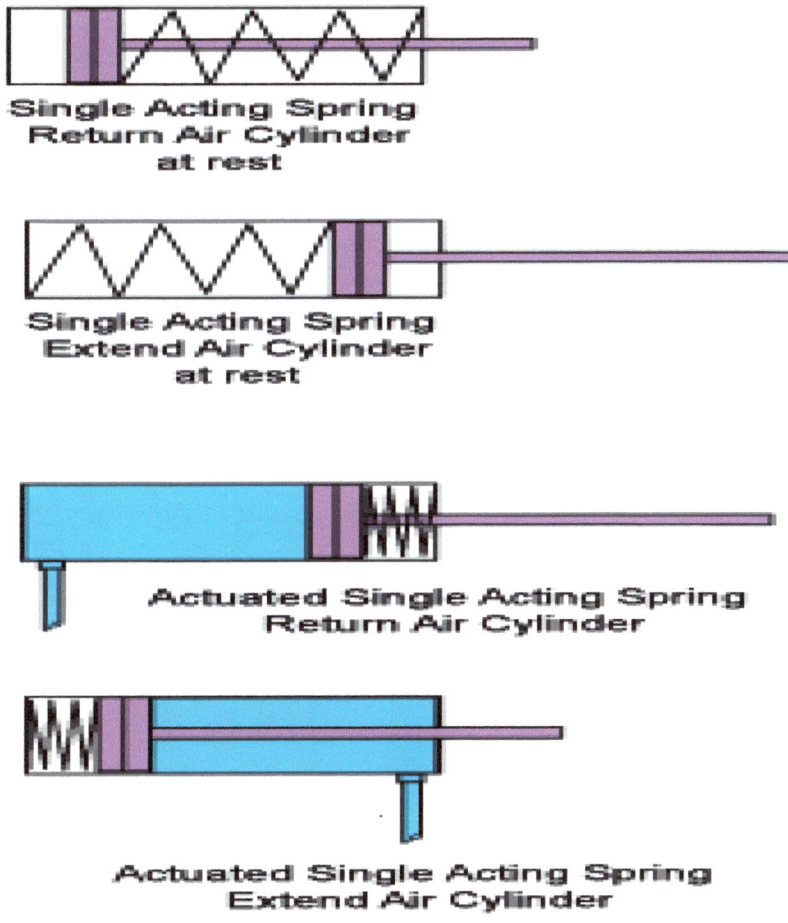

fig.4.1 type of pneumatic cylinder

4.2.2 DOUBLE ACTING CYLINDERS

Double acting Cylinders (DAC) use the force of air to move in both extends and retract strokes. They have two ports to allow air in, one for outstroke and one for in stroke and one for in stroke. The double acting cylinder as shown in the Fig.4.2. In the one stroke both extends and retraction takes place

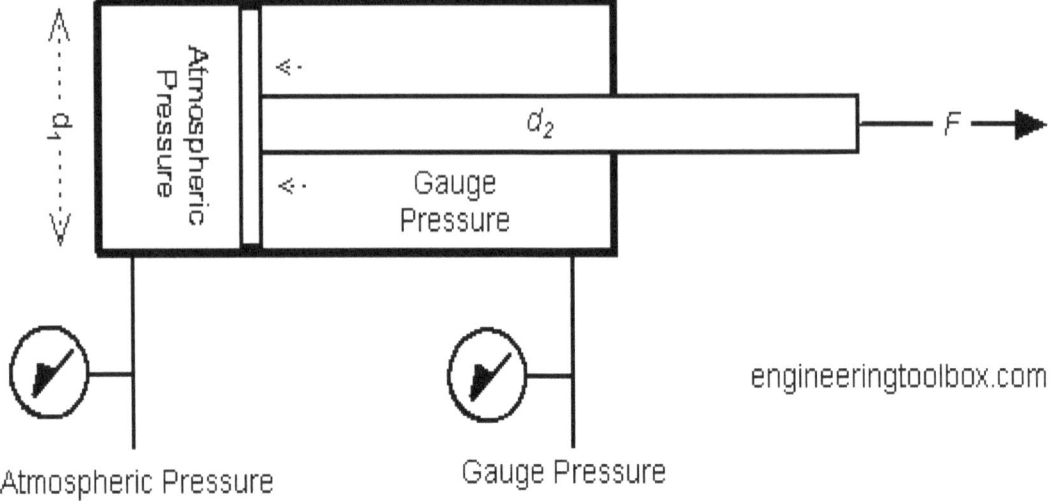

Fig.4.2double acting cylinder

4.3 OTHER TYPES

Although SACs and DACs are the most common type of pneumatic cylinder, the following type is not particularly rare. Rotary air cylinders: actuators that use air to impart a rotary motion Rod less air cylinder: these have no piston rod. They are actuators that use a mechanical or magnetic coupling to impart forces, typically to a table or other body that moves along the cylinder body, but does not extend beyond it.

4.3.1 RODLESS CYLINDER

Some rod less type has aslot in the wall of the cylinder. That slot is closed off for much of its length by two flexible metals sealing bands. The inner one prevents air from escaping, while the outer one protect the slot and inner band. The piston is actually a pair of them, part of a comparatively long assembly.

They seal to the bore and inner band at both ends of the assembly. Between the individual pistons, however, are cramming surface that "peel off" the band as the whole sliding assembly moves towards the sealed volume, and "replace" them as the assembly moves away from the other end it clearly shows in the Fig 4.3. Between the cramming surfaces is part of the moving assembly that protrudes through the slot to move the load. Of course, this means that the region where the sealing bands are not in contact is at atmospheric pressure.

Fig. 4.3 round less cylinder

Another type has (or a single cable) extending from both (or one) ends of the cylinder. The cable are jacketed in plastic (nylon, in those referred to), which provides a smooth that permits sealing the cables where they pass through the ends of the cylinder. Of course, a single cable has to be kept in tension.

Still others have magnets inside the cylinder, part of the piston assembly, that pull along magnets outside the cylinder wall. The latter are carried by the actuator that moves the load. The cylinder wall is thin, to ensure that the inner and outer magnets are near each other. Multiple modern high-flux magnet groups transmit force without disengaging or excessive resilience.

4.4. SIZES

Air cylinders are available in a variety of sizes and an typically range from a small 2.5mm air cylinder, which might be used for picking up a small transistor or other electronic component, to 400 mm diameter air cylinders which would impart enough force to lift a car. Some pneumatic cylinders reach 1000 mm in diameter, and are used in place of hydraulic cylinders for special circumstances where leaking hydraulic oil could impose and extreme hazard.

4.5 MATERIALS

The pneumatic cylinders designed for educational use typically have transparent outer sleeves (often Plexiglas), so students can see the piston moving inside.

The pneumatic cylinders designed for clean room applications often use lubricant-free Pyrex glass pistons sliding inside graphite sleeves.

4.6 PNEUMATIC CYLINDERS

Cylinder1

The double acting pneumatic cylinder as shown in the Fig. 4.5The stroke length of the cylinder is 230mm and the cylinder is made up of stainless steel material. The diameter of the cylinder is 65mm shaft diameter is 20mm. The pneumatic cylindersare given in the Table: 4.1

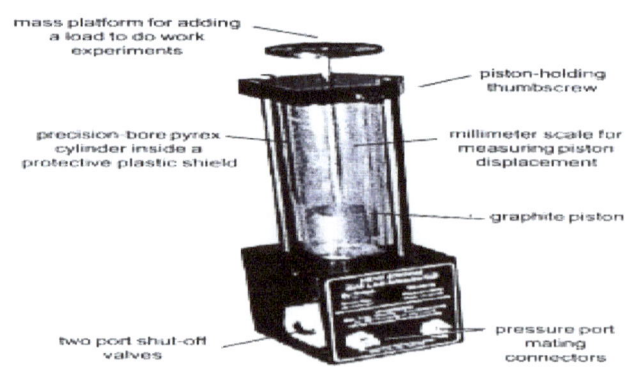

Figure 1. Base apparatus

fig.4.4pneumatic cylinder

Name	Dimension
Stroke length	230mm
Diameter of the cylinder	65mm
Diameter of the shaft	20mm

Table 4.1 specification of cylinder 1

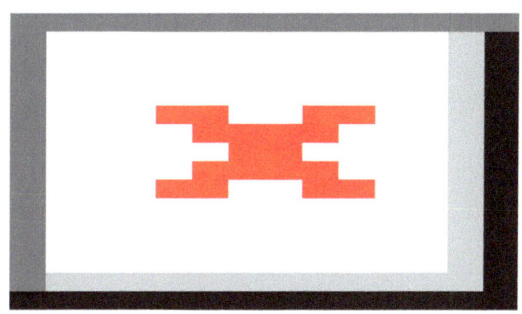

Fig.4.5 pneumatic cylinder

Cylinder2

The double acting pneumatic cylinder as shown in the Fig.4.6. The stroke length of the cylinder is 110mm and the cylinder is made up of stainless steel material. the diameter of the cylinder is 60mm shaft diameter is 15mm. The pneumatic cylinders are given in the Table: 42

Name	Dimension
Stroke length	110mm
Diameter of the cylinder	60mm
Diameter of the shaft	15mm

Table 4.2 specification of cylinder 2

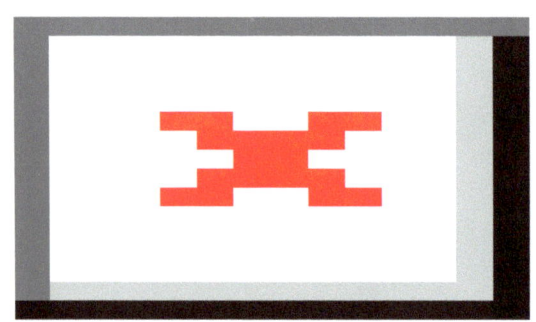

Fig. 4.6 pneumatic cylinder

4.7 FLOW CONTROL VALVE

A flow control valve regulates the flow or pressure of a fluid. Control valves normally respond to signals generated by independent devices suchas flow meters or temperature gauges.

Control valves are normally fitted with actuators and positioners. Pneumatically-actuated globe valves and Diaphragm Valves are widely used for control purposes in many industries, although quarter-turn types such as (modified) ball, gate and butterfly valves are also used.

Fig.4.7flow control valve

Control valves can also work with hydraulic actuators (also known as hydraulic pilots). These types of valves are also known as Automatic Control Valves. The hydraulic actuators will respond to changes of pressure or flow and will open/close the valve. Automatic Control Valves do not require an external power source, meaning that the fluid pressure is enough to open and close the valve. The automatic flow control valve is shown in the Fig.4.7.

Automatic control valves include: pressure reducing valves, flow control valves, back-pressure sustaining valves, altitude valves, and relief valves. An altitude valve controls the level of a tank. The altitude valve will remain open while the tank is not full and it will close when the tanks reaches its maximum level. The openings and closing of the valve requires no external power source (electric, pneumatic, or man power), it is done automatically, hence its name.

4.7.1 SYMBOL

Fig. 4.8 flow control valve symbol

The symbol for the adjustable flow control valve as shown in the Fig.4.8. A device that controls the flow of pressurized air in a pneumatic circuit. Speed control valves are used to adjust the rate of airflow into or out of a pneumatic circuit or component. The rate of flow through a circuit or component affects the speed of the component. The higher the flow rate, the faster the component will operate. Note: Controlling air flow out of the cylinder is the preferred choice of accurate and smooth control of slower moving actuators.

Use a bicycle pump to fill the pneumatic reservoir. Do not pressurize the storage reservoir to more than 150psi, and never construct or use a pneumatic circuit without a regulator. This will greatly reduce the chances of over pressurizing the system. It is not necessary to generate storage tank pressures greater the 150 psi.

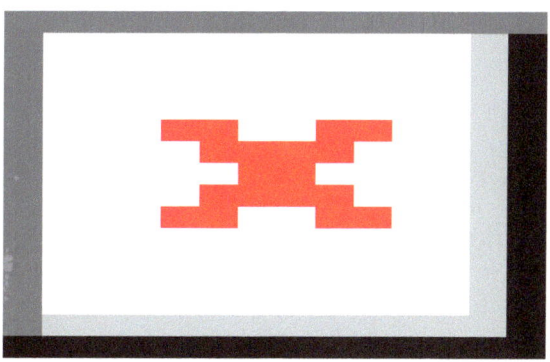

Fig. 4.9 flow control valve pro-e design

4.8 PNEUMATIC SWITCHES

It is used to control the flow of air by electrical signal. Here it has two switches. The positive end is directly connected to the supply .The another end is connected through NO (NORMAL OPEN) of the pneumatic switch. In two switches one is for top cylinder another one is for bottom cylinder.

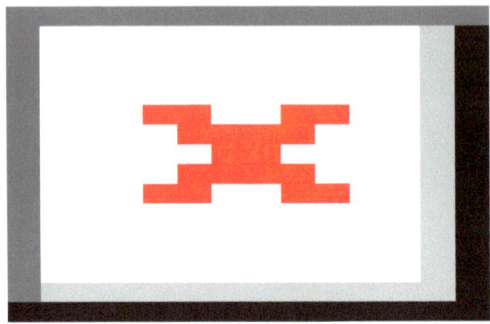

fig. 4.10 pneumatic switch-1

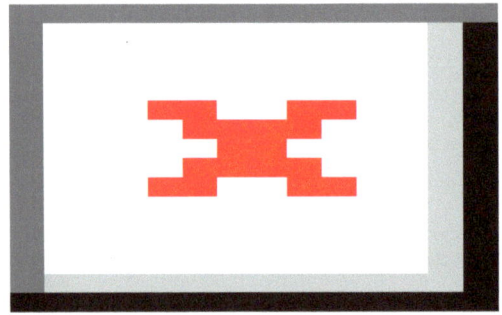

Fig. 4.11 pneumatic switch-1

4.9 TUBES

It is used to connect the cylinder and the control valves.

4.10SOLINOIDVALVES:

The signal from the switch is send to the pneumatic valves.

The valve controls the flow direction of air.

Here two valves are used to control the flow one is for upper cylinder another one is for lower cylinder.

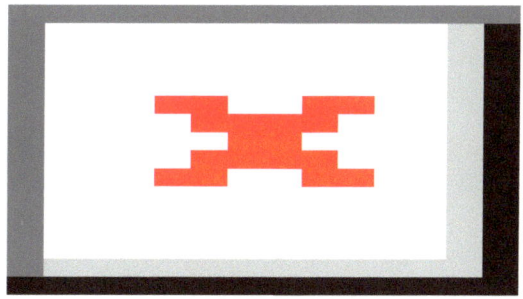

Fig.4.12 solenoidvalve pro-e design

4.11 TOOL

- It is used to de-husk the coconut.

- So it must be sharp and strong.

- It is made up of cast steel.

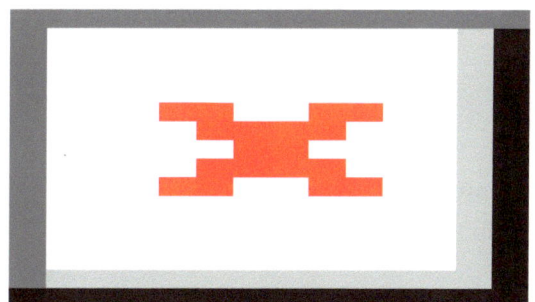

Fig.4.13 Tool

4.12 FRAME

- It is body of the machine.

- It is made up of cast steel.

- 320x25x25mm L angle of 6mm thickness & 465x25x25mm L angle of 6mm thickness are used for construction of frame.

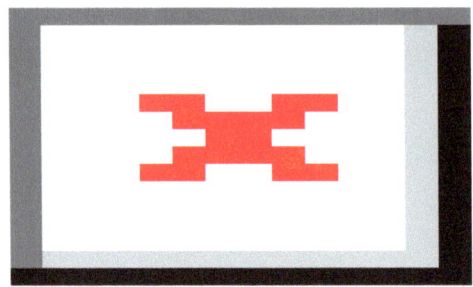

Fig.4.14 320 L-Angle

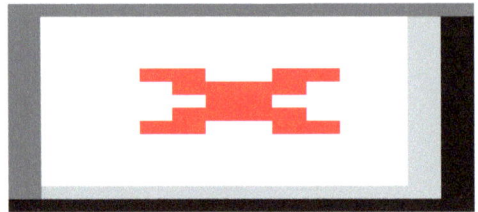

Fig.4.15 465 L-Angle

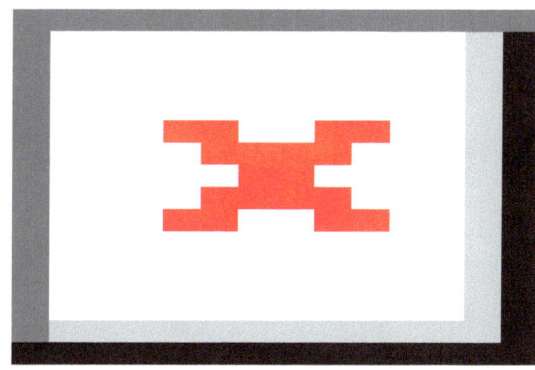

Fig.4.16 Total Frame

4.13 COCONUT SEATING PLATE:

- It is used to carry the coconut.

- It is in round shape.it is made up of cast steel.

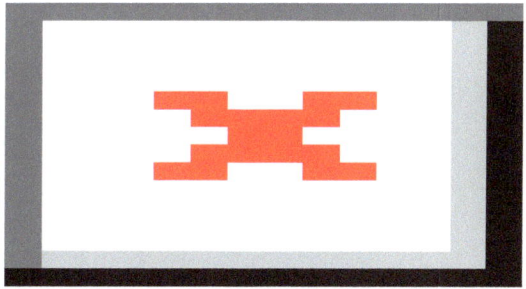

Fig.4.17 Coconut seating plate

4.14 CYLINDER CLAMPING PLATE

It is used to hold the cylinder in the machine here there are two plates are used.

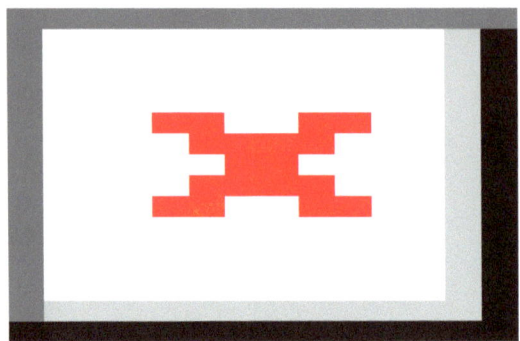

Fig.4.18 Upper cylinder clamping plate

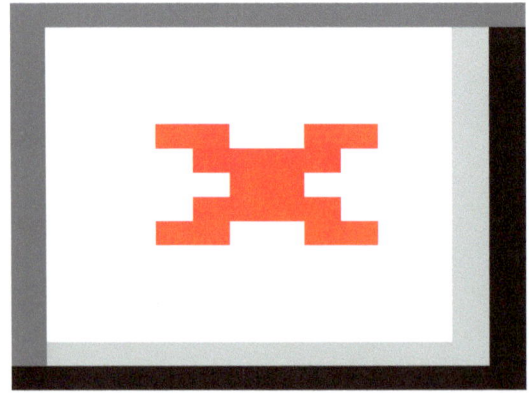

Fig.4.19Lower cylinder clamping plate

4.15 BUSH

It is used to connect the tool with the connecting rod the bush is shown in the fig.3.15. It is made up of cast steel.

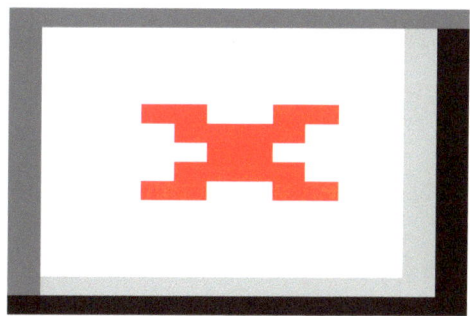

Fig.4.20 Bush pro-e design

4.16 CONNECTING ROD

- It is used to connect the tool with the bush.

- It is made up of cast steel.

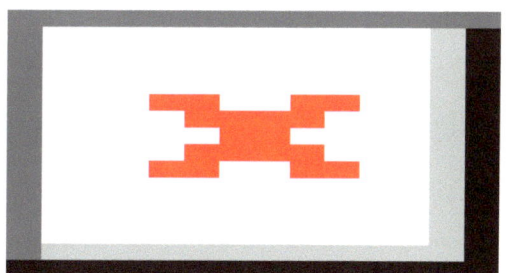

Fig 4.21.Connecting Rod

CHAPTER 5

WORKING OF COCONUT DE-HUSKING MACHINE

Coconut de-husking machine is based on the pneumatic power.it consist of pneumatic cylinder, tool, coconut seating plate, frame, tube, pneumatic switches, pneumatic valve, control valve ,cylinder bush.

The de-husking is started when we operate the pneumatic switches.

5.1 SWITCH CONNECTION

Here it has two swiches.The positive end is directly connected .the another end is connected through NO (NORMAL OPEN) of the pneumatic switch. In two switches one is for top cylinder another one is for bottom cylinder.

5.2 CYLINDER OPERATION

Here two cylinders are used one is for vertical movement (bottom cylinder), another one is for expansion of tool.

5.2.1 Top cylinder operation:

The initial position of the tool is not expanded position. Here the tools have sharp edge. When the cylinder shaft moves down wards the tool will get expanded.

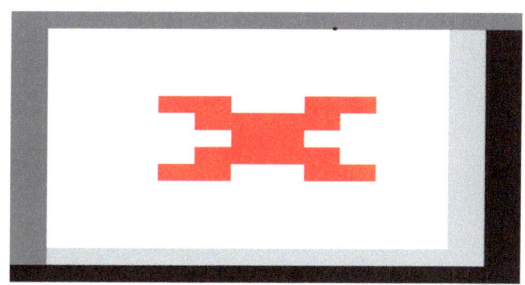

Fig.5.1 Tool position 1

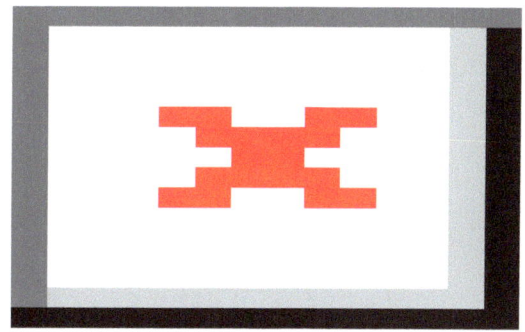

Fig.5.2 Tool position 2

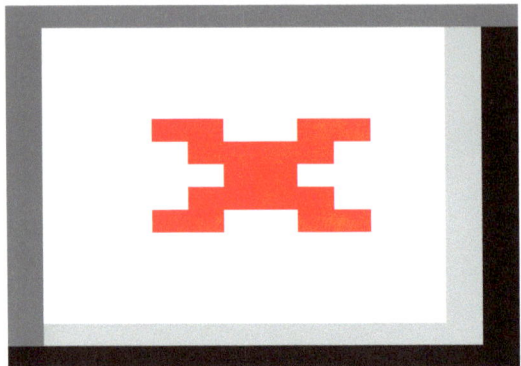

Fig.5.3 Tool shape

5.2.2 Bottom cylinder operation:

- Bottom cylinder is first in non-expanded position the coconut plate is in bottom.
- When the cylinder move upward coconut plate is also move upward.
- When the cylinder moves upward it forces the coconut towards the tool.

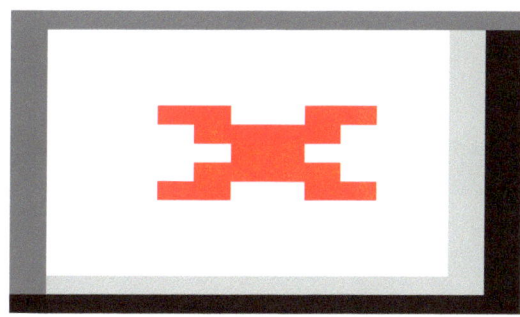

Fig.5.4 Initial Position 2nd Cylinder

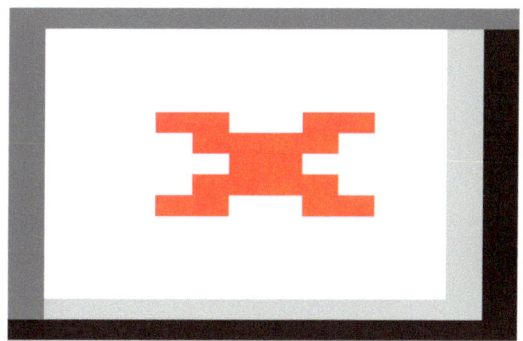

Fig.5.5 Final Position 2nd Cylinder

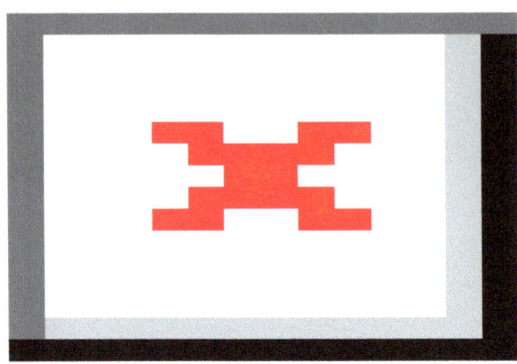

Fig.5.6 Shape of coconut seating Plate

CHAPTER 6

CALCULATION

6.1 PNEUMATIC CYLINDERS

6.1.1 Pressure, radius, force relationships

Although the diameter of the piston and the force exerted by a cylinder are related, they are not directly proportional to one another. Additionally the typical mathematical relationship between the two assume that the air supply does not become saturated. Due to the effective cross sectional area reduced by the area of the piston rod, the in stroke force is less than the outstroke force when both are powered pneumatically and by same supply of compressed gas.

The relationship, between force on outstroke, pressure and radius, is follows:

$$F = p(\pi r^2)$$

This is derived from the relationship, between force, pressure and effective cross-section, which is

F=Pa

With the same symbolic notation of variables as above, but also a represents the effective cross-sectional area.

On in stroke, the same relationship between force exerted, pressure and effective cross sectional area applies as discussed above for outstroke. However, since the cross sectional area is less than the piston area the relationship between force, pressure and radius is different. The calculation isn't more complicated through, since the effective cross sectional area is merely that of the piston less that of the piston rod.

For in stroke, therefore, the relationship between forces exerted, pressure, radius, of the piston, and radius of the piston rod is as follows.

$$F= p(\pi r^2 - \pi R^2)$$

Where

    F represents the force exerted

r represented the radius of the piston

    R represented the radius of the piston rod

π is approximately equal to 3.14.

6.2 CALCULATION OF DOUBLE ACTING CYLINDER:

Double Acting Cylinder

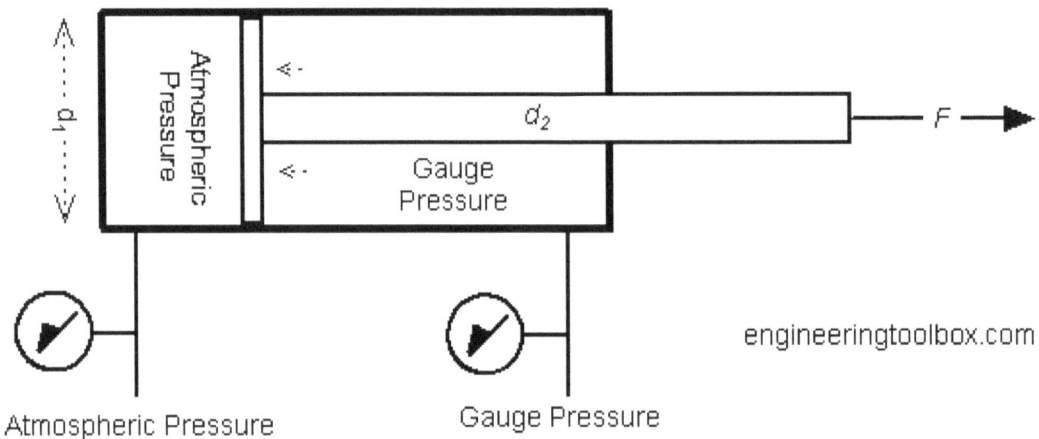

Fig. 6.1 Double Acting Cylinder

The force exerted by double acting pneumatic cylinder on outstroke can be expressed as (1). The force exerted on in stroke can be expressed as

$$F = p \, \pi \, (d_1^2 - d_2^2) / 4 \qquad (2)$$

Where

d_1 = full bore piston diameter (m)

d_2 = piston rod diameter (m)

Calculate Force when Pressure and Diameter are known:

Cylinder 1

Pressure (bar) =30 bar

Cylinder Diameter (mm) =60mm

Piston Rod Diameter (mm)=15mm

Example - Double Acting Piston

The force exerted from a single acting pneumatic cylinder with 1 bar (10^5 N/m²), full bore diameter of 60 mm (0.06 m) and rod diameter 15 mm (0.015 m) can be calculated as

$F = p \pi (d_1^2 - d_2^{2)} / 4$

$= (30 \times 10^5 \text{ N/m}^2) \pi [(0.06 \text{ m})^2 - (0.015 \text{ m})^2] / 4$

= <u>7952.15</u> N

= <u>7.9521</u> kN

- Instroke capacity is reduced compared to outstroke capacity - due to the rod and reduced active pressurized areal

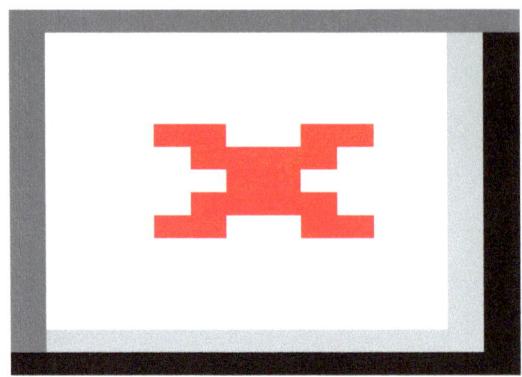

Fig 6.2 Calculation of cylinder 1

Cylinder 2

Pressure (bar) =30 bar

Cylinder Diameter (mm) =65mm

Piston Rod Diameter (mm)=20mm

Example - Double Acting Piston

The force exerted from a single acting pneumatic cylinder with 1 bar (10^5 N/m^2), full bore diameter of 60 mm (0.06 m) and rod diameter 15 mm (0.015 m) can be calculated as

$F = p \, \pi \, (d_1{}^2 - d_2{}^{2)} / 4$

$= (30 \times 10^5$ N/m$^2) \, \pi \, [(0.065$ m$)^2 - (0.02$ m$)^2] / 4$

= <u>9012.44N</u>

= <u>9.0124</u> kN

- Instroke capacity is reduced compared to outstroke capacity - due to the rod and reduced active pressurized areal

Fig 6.3 Calculation of cylinder 2

6.3 CALCULATION OF VOLUME:

Cylinder 1

Cylinder Diameter (mm) d_1=60mm

Piston Rod Diameter (mm) d_2 =15mm

Stroke length (mm) L =110mm

Cross section area (mm) = $\pi (d_1{}^2 - d_2{}^2) / 4$

35

$A = \pi \, (0.06^2 - 0.015^2) / 4$

$A = 2.6507 \times 10^{-3} \, m^2$

Volume V = area x length

$V = 2.6507 \times 10^{-3} \times 110$

$V = 0.2915 m^3$

Cylinder 2

Cylinder Diameter (mm) $d_1 = 65mm$

Piston Rod Diameter (mm) $d_2 = 20mm$

Stroke length (mm) L = 230mm

Cross section area (mm) $= \pi \, (d_1^2 - d_2^2) / 4$

$A = \pi \, (0.065^2 - 0.02^2) / 4$

$A = 3.0041 \times 10^{-3} \, m^2$

Volume V = area x length

$V = 3.0041 \times 10^{-3} \times 230$

$V = 0.6909 \, m^3$

CHAPTER 7

DESIGN OF CAMPONENTS

7.1 SOFTWARE USED FOR DESIGNING

7.1.1 Pro-e

PRO-E is the process of creation, modification and analysis of an engineering drawing by using computer. PRO-E is a computer aided drawing software.

7.1.2 Engineering Design

Cero Elements/Pro offers a range of tools to enable the generation of a complete digital representation of the product being designed. In addition to the general geometry tools there is also the ability to generate geometry of other integrated design disciplines such as industrial and standard pipe work and complete wiring definitions. Tools are also available to support collaborative development.

A number of concept design tools that provide up-front Industrial Design concepts can then be used in the downstream process of engineering the product. These range from conceptual Industrial design sketches, reverse engineering with point cloud data and comprehensive free-form surface tools.

7.1.3 Analysis

Creo Elements/Pro has numerous analysis tools available and covers thermal, static, dynamic and fatigue finite element analysis along with other tools all designed to help with the development of the product. These tools include human factors, manufacturing tolerance, mould flow and design optimization. The design optimization can be used at a geometry level to obtain the optimum design dimensions and in conjunction with the finite element analysis.

7.1.4 Manufacturing

By using the fundamental abilities of the software with regards to the single data source principle, it provides a rich set of tools in the manufacturing environment in the form of tooling design and simulated CNC machining and output.

Tooling options cover specialty tools for molding, die-casting and progressive tooling design.

7.2 COMPONENTS

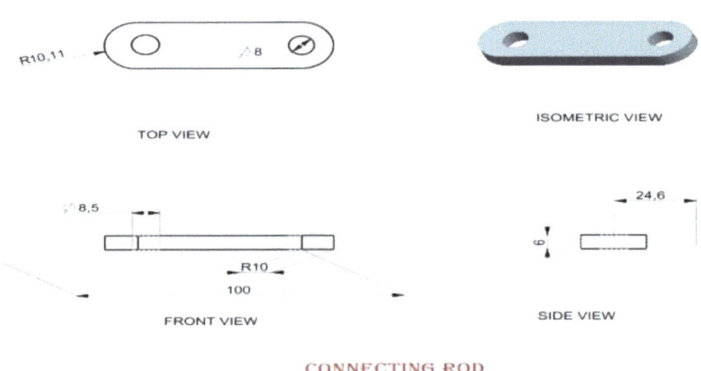

Fig .7.1 connecting rod

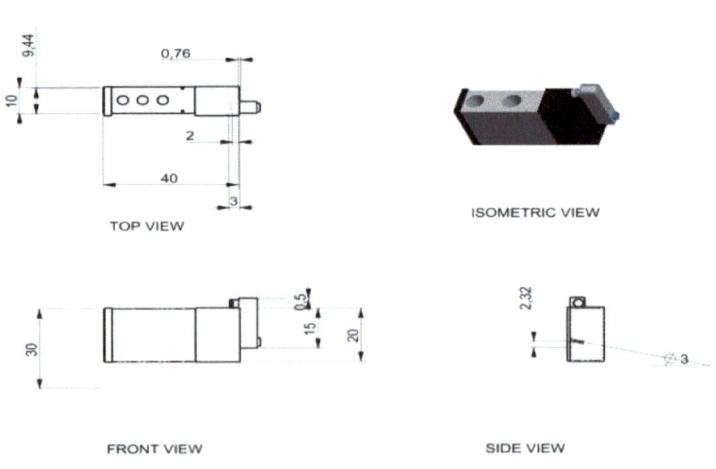

Fig 7.2 solenoid valve

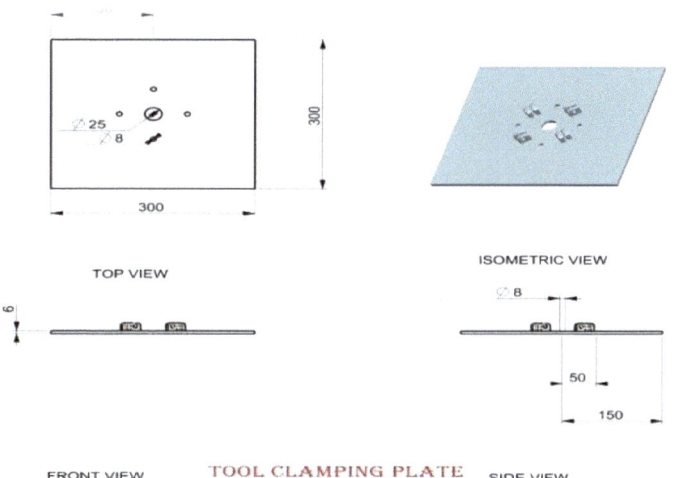

TOP VIEW

ISOMETRIC VIEW

300

300

Ø 25

Ø 8

6

Ø 8

50

150

FRONT VIEW   TOOL CLAMPING PLATE   SIDE VIEW

Fig 7.3 Clamping plate

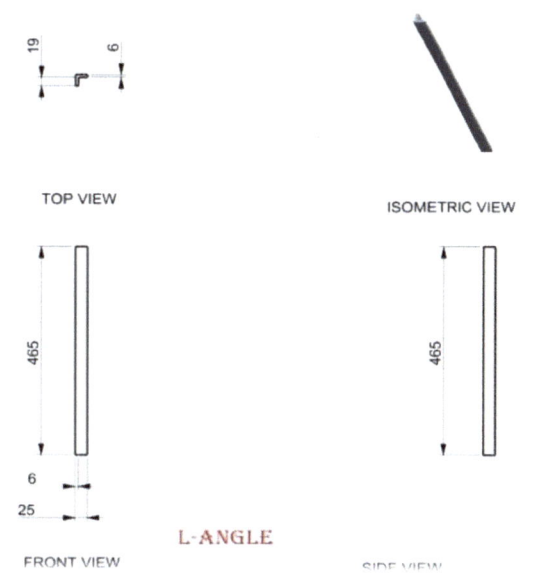

19

6

TOP VIEW

ISOMETRIC VIEW

465

465

6

25

L-ANGLE

FRONT VIEW

SIDE VIEW

39

Fig. 7.4 L-Angle

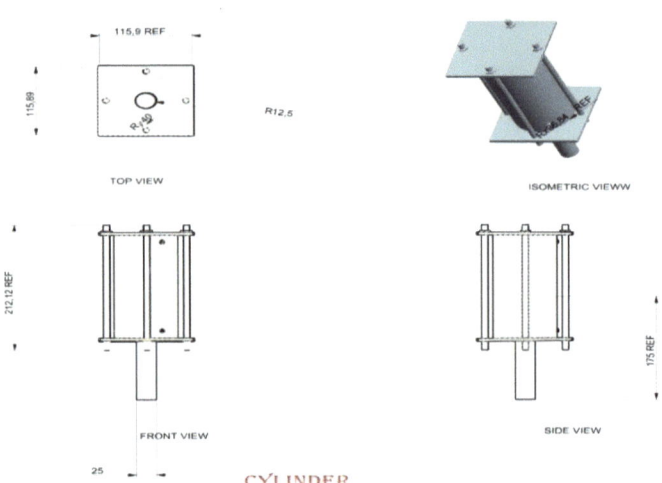

TOP VIEW

ISOMETRIC VIEWW

FRONT VIEW

SIDE VIEW

CYLINDER

Fig 7.5 cylinder

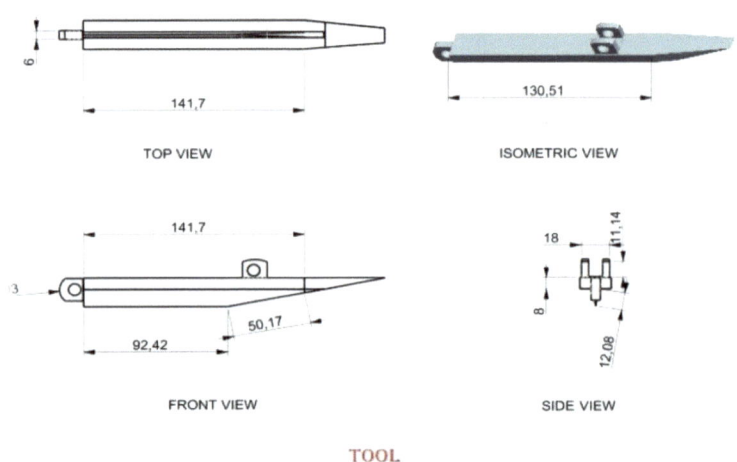

TOP VIEW

ISOMETRIC VIEW

FRONT VIEW

SIDE VIEW

TOOL

Fig 7.6 Tool

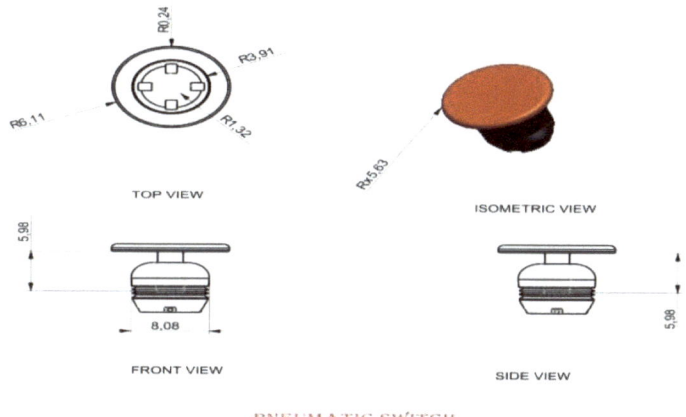

Fig 7.7 pneumatic switch

Fig 7.8 Coconut Clamping Plate

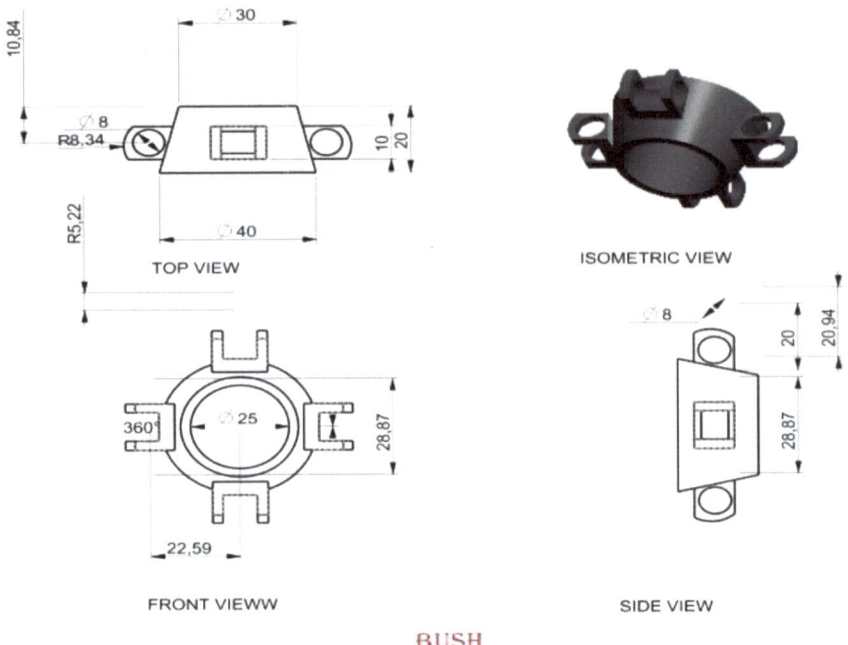

Fig 7.9 Bush

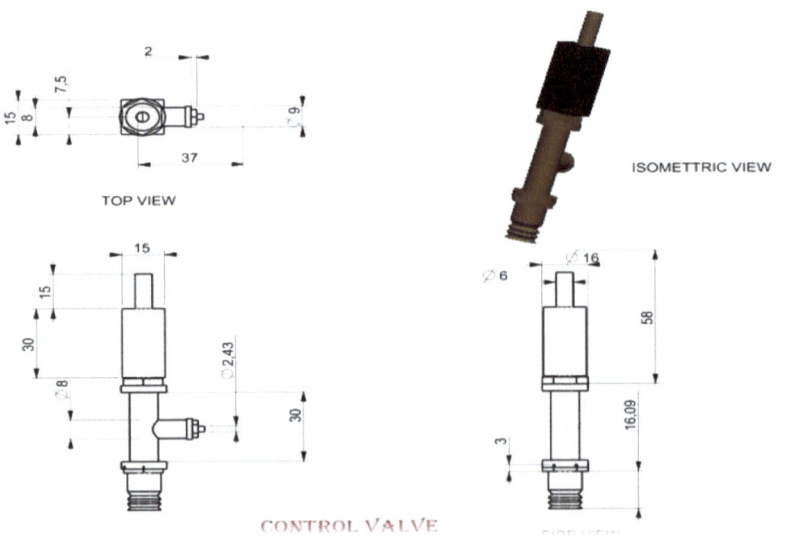

Fig 7.10 control valve

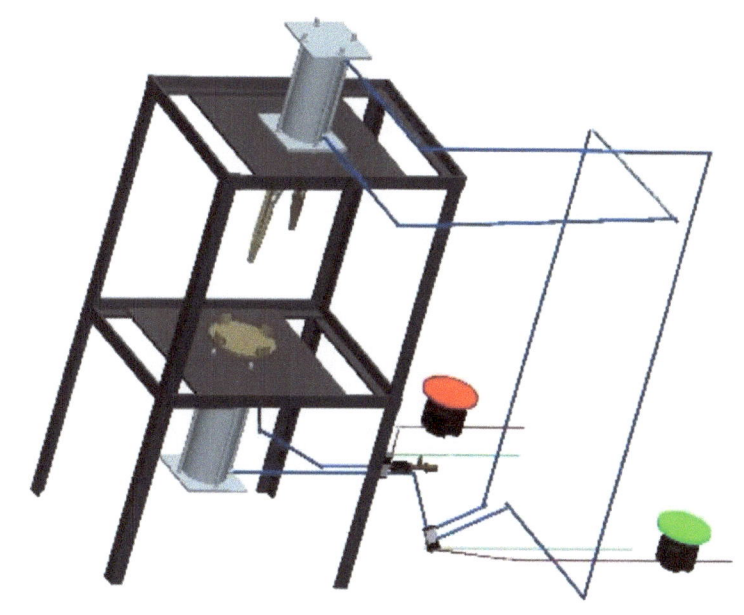

ASSEMBLED VIEW OF COCONOT DE-HUSKING
MACHINE

Fig.7.11 Assembled (isometric view) view of coconut de-husking machine

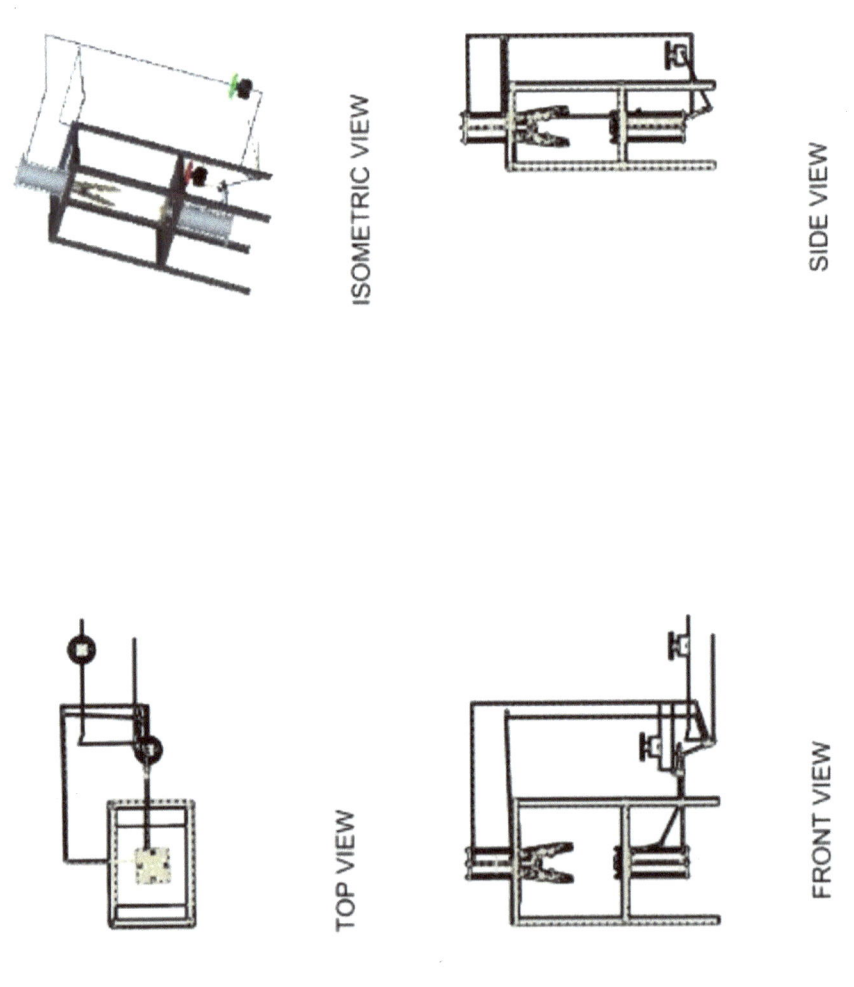

Fig.7.12 Assembled view of coconut de-husking machine

CHAPTER 8
COST ESTIMATION

Description of component	Cost (Rs.)
Pneumatic cylinders	2150.00
L-Angle	1200.00
Steel plates	300.00
Pneumatic switches	130.00
Control valve	100.00
Tubes	120.00
Solenoid valves	100.00
Wires and others	100.00
Labour cost	150.00
TOTAL COST	**4750.00**

Table 8.1 Cost estimation

CHAPTER 9

EXPERIMENTAL WORK

9.1 TESTING OF COCONUT DE-HUSKING PROCESS

The upper cylinder shaft length is constant (stroke length) $L_S = 110$ mm

The distance of the tool from connecting rod attached to end D_a $= 120$mm

The tool length is constant $L_T = 230$ mm

The distance of the tool from the center is constant $D_c = 100$mm

9.2 FACTORS INFLUENCES ON COCONUT DE-HUSKING RATE

There are two factors influences on coconut de-husking rate.

They are,

- Length of the connecting rod
- pressure

9.2.1 Factor 1 - Length of the connecting rod

Here we use different length connecting rods

We can increase the force by reducing the length of the connecting rod

Because thrust= force x distance

T1 = T2

If T1= F1 X 12

T2 = F2 X 10

F1 X 12 = F2 X 10

 F2=1.2 F1

So force can be increased by reducing the length of the connecting rod L_{cr}

9.2.2 Factor 2-Pressure

By increasing the pressure we can get high de-husking rate

When the pressure is increased the force is also increased

Because

F= P /A

F α p

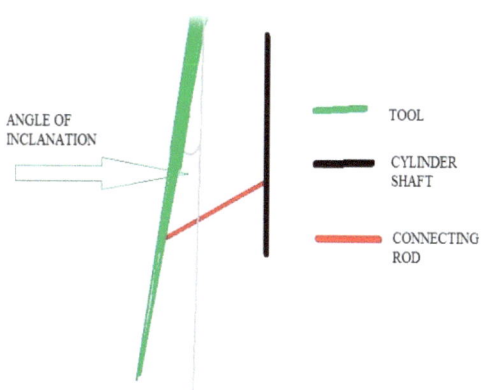

Fig. 9.1 position of tool, connecting rod, cylinder shaft

9.3 EXPERIMENTS

Ex .no	The upper cylinder shaft length is constant (stroke length) L$_S$ In mm	The distance of the tool from connecting rod attached to end D$_a$ In mm	The tool length is constant L$_T$ in mm	The distance of the tool from the center is constant D$_c$ in mm	length of the connecting rod L$_{cr}$ in mm	Tool angle In degree in mm	Pressure In bar	result
1	110	120	230	100	120	100	1bar	De-husking rate is very low
2	110	120	230	100	120	100	2bar	De-husking rate is low
3	110	120	230	100	110	95	4 bar	De-husking rate is low
4	110	120	230	100	100	95	6bar	De-husking rate is low
5	110	120	230	100	100	90	8 bar	De-husking rate is high

Table 9.1 experimental work

9.4 PHOTOGRAPHS

Fig.9.2 Assembled view of coconut de-husking machine

Fig.9.3 bottom cylinder

Fig.9.4 Top cylinder

Fig.9.4 Pneumatic switches

Fig.9.5 coconut seating plate

Fig.9.6 Tool

CHAPTER 10

CONCLUSION

This project is made with pre planning, that it provide flexibility in operation. Here we innovate the pneumatic coconut de- husking machine it is more comfortable than old methods. The operation is easier than other method. If we use this machine for mass production it is more economical. If We connect the more than one machine as parallel with a single compressor the de-husking process will be more quicker than other conventional manual method connect .when the de-husking time is reduced automatically it reduces the labor cost.

Now de-husking process is semiautomatic type it need some manual work like placing the coconut. But in future we decided to automate the coconut placing operation for de-husking process.it needs another one additional pneumatic cylinder.

www.ingramcontent.com/pod-product-compliance
Lightning Source LLC
Chambersburg PA
CBHW050817180526
45159CB00004B/1704